Rodrigo Larenas

MEGA
EARTHQUAKE & TSUNAMI
ON THE PACIFIC COASTS

Chilean experiences for Pacific Rim residents

ISBN: 1451596030
ISBN-13: 9781451596038

MEGA EARTHQUAKE AND TSUNAMI ON THE PACIFIC COASTS

Chilean Experiences for Pacific Rim Residents

Preface	5
My Experience Of The 2010 Mega Earthquake	11
What Happened In A More Technical But Simple Way?	31
Lessons For The Future	61
Complementary Information & Acknowledgments	81, 85

MEGA
EARTHQUAKE & TSUNAMI
ON THE PACIFIC COASTS

Chilean experiences for Pacific Rim residents

To all the people who suffered from this devastating experience, for those many unknown heroes who saved the lives of others and for those who are willing to prepare their communities for future shocking events

PREFACE

The mega earthquake occurred on February 27 of 2010, along the coast of the central Chile in South America; it produced an early and shocking wake-up call for millions of people. Registering 8.8 on the Richter scale, the event hit the central zone at 3:34 A.M.

This earthquake is the fifth most intense ever measured; it was so strong that *Scientific American* published the following in its March 4th newsletter: "The magnitude 8.8 earthquake that jolted Chile on Saturday was felt as far away as São Paulo. But NASA scientists are proposing that its repercussions are truly global in a geophysical sense: it likely shifted Earth's axis by about eight centimeters."

Unfortunately the tragedy didn't stop with the earthquake. As if the initial damage wasn't enough, the earthquake was followed by a devastating and also lethal tsunami that was strong enough to travel across the Pacific Ocean and hit Asia only twenty four hours later. It hit the Chilean coast with four waves, the first arriving just a few minutes after the earthquake and the last, several hours later. This tsunami was so strong that at least one wave was twenty meters or sixty-five feet high.

Before continuing I would like to prevent lecturers from falling into the temptation of thinking that these huge natural events only happen in faraway lands and affect people quite different from us. The Indonesian tsunami of December 2004 was very distant from Chile not only in miles but also in cultural terms. But now, it's clear that Indonesian experience helped to prevent more fatalities in Chile.

This book was written to urge people living in the Pacific Rim to take some time to analyze the situation in their community and take action to prevent or reduce the potential for future damage and casualties.

To understand why it is so important to be prepared, let me resume explaining what a megaearthquake like this one means. In human terms, it hit 75 percent of the Chilean population. In a country with a population of sixteen million, 800,000 people were harmed and 190,000 homes were damaged. Of the 130 hospitals located in the area that suffered the most damage, twenty-five were severely damaged, and fifty-four need some level of repair.

From an economic perspective, the impact is sobering. Damage is estimated at a cost of US$ 30,000 million or 11 percent of the country's capital stock. The infrastructure was also hurt; there are ten large bridges, 4,200 fishing boats, 420 rural water piping systems, and twenty-seven fishing coves that were severely damaged. Significant damage was also done to ports, highways, defense capabilities, energy, telecommunications, and historic structures. One little-known fact: 100 million liters of wine were lost.

Unfortunately, one thing that we can be sure of is that earthquakes and their consequences will continue to occur. The idea of this book is to share experiences and thoughts in the hope that at least some of them may help people improve their preparation for events like this.

Because of the timing of a natural disaster is never a certainty, it's my impression that people have a tendency to forget and dismiss dangerous situations. This is especially true when there is an interval of more than a few years between events; people grow complacent because the disaster hasn't happened in such a long time. Examples of this mindset are easy to see with edifications built over riverbeds that once flooded, etc. It's my opinion that we need to couple a *normal* life with the safety conditions necessary to minimize damage and casualties whenever we face such natural disasters.

It's also my belief that there are some aspects of the construction code in use in Chile that helped to contain and limit the damage of the earthquake; it is not perfect and there were not only structural but also personal casualties.

In the book, readers will start on chapter 1 by realizing that a scientist predicted the event, followed by a brief description of my experience—it's a personal but accurate view of the event.

Then, in chapter 2, readers will find an explanation of what happened from different perspectives. One describes the earthquake and the tsunami with a more technical orientation, another describes key geological aspects that are related to these to this natural event, and a third seeks to understand it from an economic and cost perspective. The chapter includes references that are easy for nonscientific people to understand.

In chapter 3, readers will have my view of different aspects of real life under the stress of a megaearthquake; some behaved in a good way and some in a bad way. It's concrete and easy to understand; even though it may describe some problems, it's not my intention to be a critic. My goal is to present this information as a reference for helping other people improve their preparation for situations like this.

Finally, I have provided some complementary and interesting references for people that may be interested in obtaining a deeper understanding of these kind of events.

Rodrigo Larenas
Viña del Mar, Chile
March 2010

CHAPTER 1

My experience with the Megaearthquake of 2010

Science Serving Humanity

30 March 2007, Inter-seismic strain accumulation measured by GPS in the seismic gap between Constitucion and Concepcion in Chile article is received by Elsevier (www.elsevier.com/locate/pepi).

I highlight two important sentences from the abstract of this article:

a) The Concepcion-Constitucion area (35-37°S) in southern central Chile is very likely a mature seismic gap.

b) Therefore, in a worst scenario, the area already has a potential for an earthquake of magnitude as large as 8 to 8.5, should it happen in the near future.

This article was accepted on February 10, 2008 and prepared by the following scientists: J.C. Ruegg, A. Rudolff, C. Vigny, R. Madariaga, J.B. Chabalier, J. Campos, E. Kausel, S. Barrientos, and D. Dimitrov.

Terror Phase 1

Viña del Mar, Chile, Saturday February 27th, at 3:54 A.M. everybody was sleeping at home. Suddenly a very strong and deep noise from the underground anticipates the imminent arrival of a huge earthquake. I jumped out of my bed and ran upstairs shouting to wake up my four children and take them out the house.

Once upstairs, I grabbed my youngest, my five-year-old son, and went for my eight- and twelve-year-old daughters, while checking that my thirteen-year-old son was following my instruction and running downstairs toward the main entrance.

Meanwhile, the earthquake was getting stronger and stronger; every moment it was more and more difficult to walk. My wife, who is six months pregnant, moved to the main entrance and by the time we all got there, the earthquake was at the peak its intensity.

The situation is quite hard to describe: the way in which the house was moving, the noise of everything breaking, and the roar of the ground creates a terrifying moment that just allows no time to understand it.

My house is very old—more than sixty years. It's made with adobe and wood, and to make it worse, the soil in that part of Viña del Mar is sand. Its roof is very heavy because it has adobe tiles, and it was heavily damaged in the 1985 earthquake. At that time I was fourteen-years-old, so I could remember being in that situation and the danger to which we were exposed.

But the movement was so strong that we were sitting on the floor trying to hold our bodies to the wall. We had no idea how long it would last. We just knew that something terrible was going on and that there was no guarantee that we would survive.

Then finally, the movement of the earth became less intense. My memory was clear that there will be other shocks, so I moved my family inside the car; I went back home to get essential things, and we moved to a safer place. My family was terrified.

Once in the house there is no light, phone service, etc. I could hear people crying and shouting in the street. The ground is still moving gently. I picked up some clothes, a Finland's made Led lantern (MICA), two twenty-liter containers of water, and a cooler with food that I had prepared years in advance in case of an emergency. As a complementary practice, my car's gas tank is always refilled when it gets to one-quarter tank, so I know I can drive for at least 100 kilometers.

I cut the gas, electricity, and water supply, and then closed the house. Then I got in the car and headed, with my family, to my grandmother's building, which is only a few blocks away. Because there is moonlight, I can see the expressions of people in the street: confusion and dread. While moving slowly, I'm trying to see if there's damage, and I quickly realize that are cracks in the streets, street lights broken, and buildings damaged.

Finally I get to my grandmother's building. Everything is dark. The only light comes from my lantern. I jumped over the two- meter-high gate. Almost all the people in that building are old; they are afraid. I get into my grandmother's apartment, and I see her and my aunt sitting on a sofa. They are very scared but okay. My grandmother is ninety-years-old. I ask her to come with me to my mother's house, which is ten kilometers away. She doesn´t accept my offer. The elevator is not working, and she doesn't want to move. I'm concerned; before I leave the building, I go upstairs again and repeat my invitation without success. I take a quick look at the structure and check that there is no evidence of structural damage.

I get into the car again. My wife is also concerned about my grandmother. I turn on the radio. There are no FM stations transmitting. I move to AM, looking for information about the earthquake's intensity and location. No one's knows anything. I'm on the way to my mother's house; my father is flying from Singapore back to Chile, so I know she is alone with my sister.

But suddenly I hear two journalists speaking. One is saying that there is no risk of tsunami. I'm getting angry because I believe that there is a definite risk of tsunami. I was in the navy for fourteen years, and I'm very certain in that if the intensity is over 7.5 and the epicenter of the earthquake is at sea, there is a high probability of this kind of event.

I get to my wife´s parents' home. I don't know what time it is, but I guess it should be around 4:30 A.M. The house is located at the top of a hill. All the people and neighbors are in the street. I leave my wife and children, and pick up my father-in-law who is also interested in going with me to check my mother's house. Because I still believe there is a high probability of a tsunami, I take a road that goes over the hills instead of the coastal road.

I get to my mother's house. Again, all the people and neighbors are in the street. Fortunately my brother arrived before me, so I know that he's okay. We check the house for damage and verify that there is no danger. We try to make phone calls, but it's impossible. We talk about the rest of the family, but we have no way to know how they are.

I return to the car again and start inspecting the buildings that my father-in-law's company has been building for more than fifteen years; this activity took several hours.

The Terror Phase 2

By 9:00 A.M. we learned that the magnitude of the earthquake had been 8.8 on the Richter scale and that it was located off-shore from the Maule region, which is a bit south of the central region of Chile. I'll explain the Richter scale later, but in order to give an idea of how intense this logarithmic scale can be, the Haiti earthquake on January 12, had a magnitude of 7.0 and resulted in 220,000 fatalities.

But what we didn't know, at least in Viña del Mar, was that just a few minutes after the earthquake, a tsunami hit the coast with several waves and such intensity that it also arrived in Hawaii fourteen hours later.

In Chile, if the earthquake was terrifying, the tsunami was devastating. Because it was dark, people who lived along coves or in houses close to the coast were struck by these waves that they didn't see. They lost not only their goods but also in more than 300 cases (preliminary estimation), their lives.

The Chilean navy has established that there were four waves of different intensity. In the following table, it's shown the area affected and the time at which they were hit by the waves:

Table I

Tsunami Arrival				
Area	**Wave 1**	**Wave 2**	**Wave 3**	**Wave 4**
VALPARAISO	04:01 (I)	04:50 (I)	05:20 (I)	05:40 (I)
J. FERNANDEZ	04:25 (A)	04:40 (A)	N.A.	N.A.
SAN ANTONIO	03:50 (A)	04:20 (A)	N.A.	N.A.
PICHILEMU	03:48 (A)	04:15 (A)	N.A.	N.A.
CONSTITUCION	03:49 (A)	04:17 (A)	04:50 (A)	05:20 (A)
TALCAHUANO/ DICHATO	03:54 (I)	05:30 (A)	06:00 (A)	06:40 (A)

Source: Chilean Navy. I: Instrumental data. A: different sources of data. N.A. not available

As people may imagine, the differences in the time of arrival of the different waves is still not understood. There are a number of scientists working on this because this understanding may help save lives in future events.

The damage caused by the tsunami was several times worse than that caused by the earthquake. It is simple: it was lethal. Nobody hit by that enraged ocean could defend themselves.

The impact of the sea was so strong that there are some cases in which people lost their parents, friends, or children literally from their arms. It didn't differentiate victims by age, social position, or sex. The waves were so destructive that made the 8.8 earthquake, looked like a minor event.

The tsunami hit the Chilean coast with great intensity, along more than 550 kilometers (312 miles). It also hit with great strength, causing fatalities at Juan Fernandez Island, which is more the 600 kilometers (323 miles) away.

But here is the worst: the tsunami we are talking about consisted not only of one wave but also of several others that came from different directions and were of different heights and energy. This is shocking to me because even though I realized the high probability of a tsunami and protected my family against it, I didn´t imagine this *lethal detail* of receiving waves from different directions over several hours.

By the way, the reason Viña del Mar didn't suffer damage from the tsunami seems to be related to several factors from which it is possible to highlight the fact that it's located on a deep bay open to the north of the epicenter.

The Country

Before continuing, it's important to learn a bit about this country to better understand the impact of these disasters relative to the Chilean population and economy; this may be especially helpful for readers that may live in large and or richer countries.

Chile is a long and thin country, located along the west coast of South America. In the north, it borders Peru, and in the east, Bolivia and Argentina. In the south it runs down to the Antarctic and in the west, it meets the Pacific Ocean.

The population is approximately sixteen million and the gross domestic product (GDP) is around US$160,000 million per year with a GDP per capita of US$10,000.

It's main industries are copper, foodstuffs, fish processing, iron and steel, wood and wood products, transportation equipment, cement, and textiles. The population below poverty is (or was) around 13 percent and the labor force is about 7.3 million people.

The area affected by the earthquake is located between Concepción and Valparaiso, including a small but beautiful island called Juan Fernandez (also known as Robinson Crusoe). This area is where an important part of the production of wood products, refineries, and foodstuffs are based.

As can be seen in the following tables, the earthquake hit the main metropolitan areas and also a large portion of other regions, affecting the vast majority of the population:

Table 2

Denomination	Region	Habitants
Gran Santiago	Metropolitana de Santiago	5.428.590
Gran Valparaíso	V de Valparaíso	803.683
De O'Higgins	VI del Libertador General Bernardo O'Higgins	866.249
Maule	VII del Maule	991.542
Gran Concepción	VIII del Biobío	666.381

Source: Wikipedia

Terror Phase III

By Sunday, we had more information about what had happened and the material damage caused by the disasters. We knew that only one building had collapsed because of the earthquake and that there might still be survivors inside. This was a bad thing because there were deaths.

In other sense, it was a good thing because only one building collapsed; this meant that the construction code was good enough to save lives by preventing the collapse of structures.

The bad and unexpected thing that really hurt Chilean society was the behavior of some small groups of people, who plundered not only supermarkets but also every kind of shop that was in the area after the catastrophe.

They literally stole goods like TVs and washing machines and in some cases were so violent that they even burned some large stores like La Polar in Concepcion.

These events forced the government to send the military to help the police and to declare curfew. Even though this may be seen a restrictive action, it was applauded by citizens who were scared of the vandals and their unchecked actions.

Many reasons could have motivated this exceptional and sad situation that I believe will be studied in detail by sociologists. From one side, many stores and supermarkets were damaged. As a complement to this fact, the roads were severely damaged in many places, making it difficult for trucks to move food and beverages necessary to reestablish the logistics that communities require.

From the other side, many houses had been wiped out by the sea, leaving people literally with nothing: no food, clothes, or goods of any kind. That situation created anxiety and concern in many people who didn't handle their concern in a proper, cooperative way.

CHAPTER 2

What Happened in a More Technical but Simple Way?

Earthquake in Simple Terms

Looking at this natural phenomenon in a more rigorous way it's possible to say that the rupture of a geological fault released energy creating seismic waves: the Nazca plate penetrated under the South American plate.

The epicenter of this earthquake was located offshore and had enough energy to displace the seabed and generate a tsunami. The location of the epicenter was only 345 kilometers (215 miles) from Santiago, the nation's capital and 115 kilometers (72 miles) from Concepción, another large city.

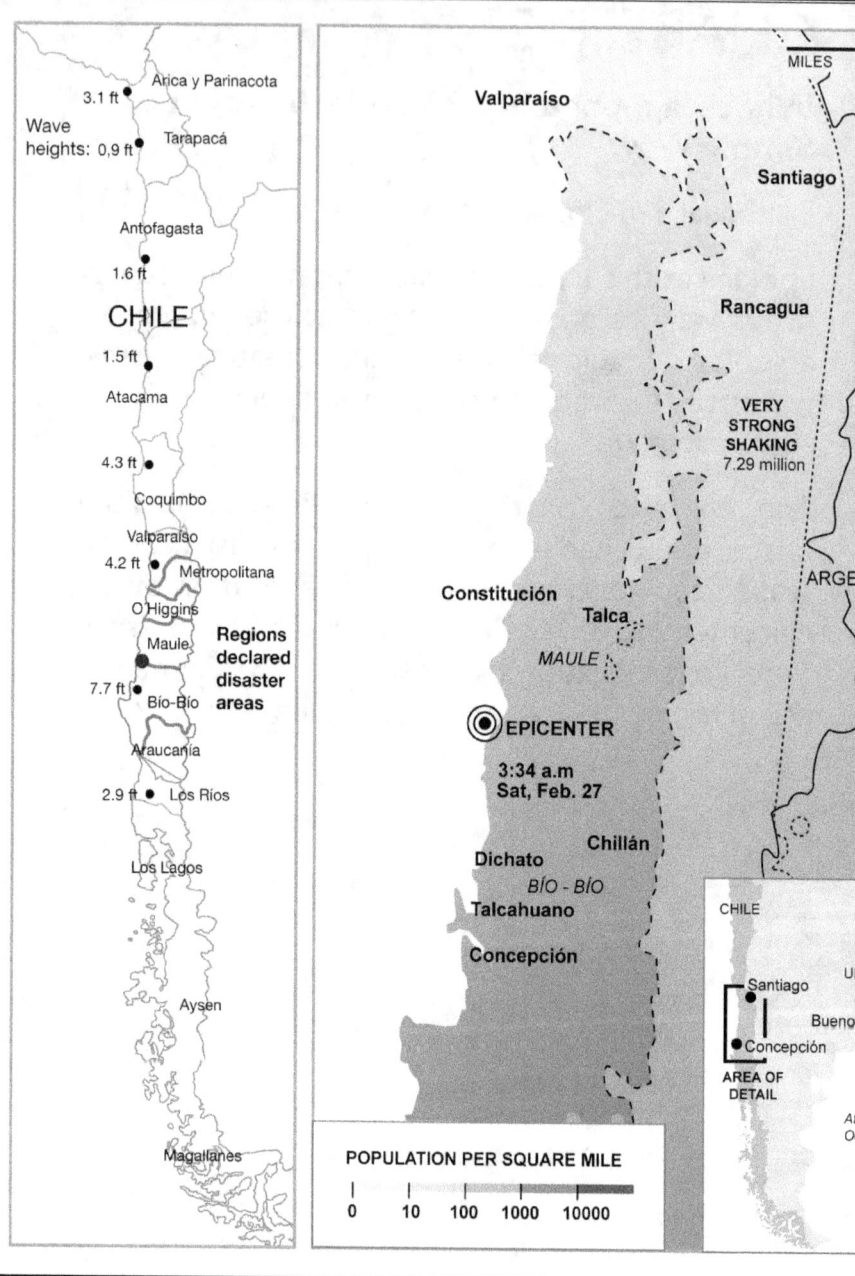

It was so intense that preliminary data indicates that Concepción moved between three to four meters.

In this moment, a question may be why do earthquakes happen? What is known as "naturally occurring earthquakes" require the accumulation of elastic strain energy. This is used to drive the fracture propagation along what it's called a fault plane.

Getting a bit deeper, there are four possible movements of tectonic plates:

1) Convergence fault: in which one plate pushes another from the underground (what happened in this megaearthquake).

2) Transformation fault: in which plates displace horizontally.

3) Divergence fault: in which plates displace moving away from each other.

4) Normal fault: in which the plates displace vertically.

In the case of convergence faults, the plate boundaries may move smoothly (also known as a-seismically) if there are no asperities. If the plates' boundaries have strong asperities, they may lock, but the continual relative movement of the plates will contribute to the accumulation of energy. Once the stress is strong enough to break the locked asperities, the energy will be released creating an earthquake.

Back to the region affected, other interesting information to take into account may be the fact that a 9.5 earthquake was recorded in 1960, only 230 kilometers (143 miles) south of this quake's epicenter. Also, since 1973, this region has been affected by thirteen earthquakes with a magnitude over 7.0.

In relation to the length of this megaearthquake and based on information recorded, even though the earthquake lasted for the "eternity" of 200 seconds (a bit more than three minutes), the U.S. Geologic Service (USGS) determined that the energy was released during the first 100 seconds (one minute and forty seconds).

Earthquake Scales

Earthquakes are typically measured using two different scales—the Richter and the Mercalli.

The Richter scale is used by seismologists to measure the size of an earthquake. The magnitude of an earthquake is determined from the logarithm of the amplitude of the waves recorded by seismographs. The Richter scale was created by 1970, replacing a scale created in 1930, and it's employed by the USGS to estimate the magnitude of every large earthquake detected by this institution.

As mentioned, the Richter scale is a logarithmic base-ten, so it´s hard to understand it with a typical linear-thinking analysis. To realize the implication of the logarithmic scale impact, the following table shows an example in which a magnitude 7.2 earthquake produces ten times more ground motion than a magnitude 6.2 earthquake, but it releases about thirty-two times more energy. The energy release best indicates the destructive power of an earthquake.

Table 3. Earthquake magnitude

Earthquake Magnitud		
Magnitude Change	Ground Motion Change (displacement)	Energy Change
1.0	10.0 times	about 32 times
0.5	3.2 times	about 5.5 times
0.3	2.0 times	about 3 times
0.1	1.3 times	about 1.4 times

(Source: USGS http://earthquake.usgs.gov/earthquakes/ eqarchives/year/eqstats.php)

Another way to understand the Richter magnitudes is shown in the following table, in which it's possible to see a relation among the magnitude, its effects, and the frequency of occurrence around the world:

Table 4: Richter Scale.

Richter magnitudes	Description	Earthquake effects	Frequency of occurrence
Magnitude, effect and frequency			
Less than 2.0	Micro	Micro earthquakes, not felt.	About 8,000 per day
2.0-2.9	Minor	Generally not felt, but recorded.	About 1,000 per day
3.0-3.9		Often felt, but rarely causes damage.	49,000 per year (est.)
4.0-4.9	Light	Noticeable shaking of indoor items, rattling noises. Significant damage unlikely.	6,200 per year (est.)
5.0-5.9	Moderate	Can cause major damage to poorly constructed buildings over small regions. At most slight damage to well-designed buildings.	800 per year
6.0-6.9	Strong	Can be destructive in areas up to about 160 km. (100 mi) across in populated areas.	120 per year
7.0-7.9	Major	Can cause serious damage over larger areas.	18 per year
8.0-8.9	Great	Can cause serious damage in areas several hundred miles across.	1 per year
9.0-9.9		Devastating in areas several thousand miles across.	1 per 20 years
10.0+	Epic	Never recorded; see below for equivalent seismic energy yield.	Extremely rare (Unknown)

Source: Wikipedia

The Mercalli scale measures the earthquake by quantifying the "perceived effects" that people have of it. Because of this, it may not be an objective way to measure the earthquake. The ranges of the scale go from I to XII as shown in the following table:

Table 5: Mercalli Scale

Mercalli Scale	
Level	**Name**
I	Instrumental
II	Feeble
III	Slight
IV	Moderate
V	Rather Strong
VI	Strong
VII	Very strong
VIII	Destructive
IX	Ruinous
X	Disastrous
XI	Very disastrous
XII	Catastrophic

Source: Wikipedia

Due to the lack of scientific rigor of this scale, and the impossibility of using it to compare the seismic intensity in a valid way, it will not be used for any purpose in this book.

Tsunami in Simple Terms

What is a tsunami? It's a wave or groups of waves of extraordinary size and energy that occur when an extraordinary event displaces vertically a huge amount of water.

37

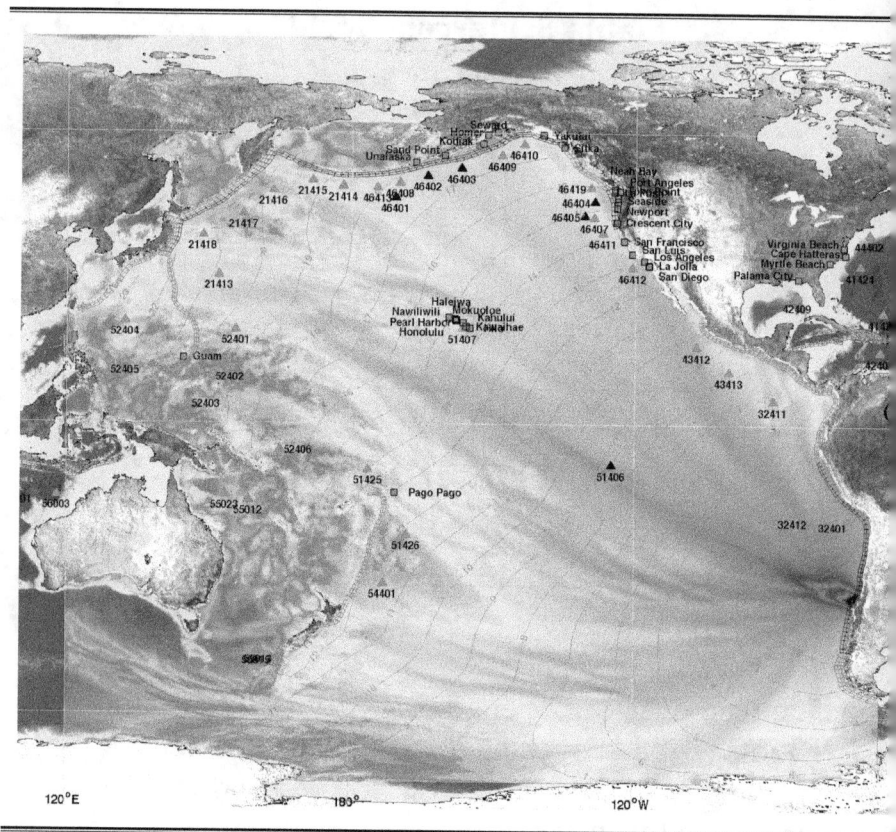

Source: NOAA / PMEL / Center for Tsunami Research

It's estimated that 90 percent of tsunamis are originated by earthquakes, but they can also be caused by volcanic activity, submarine avalanches and nuclear explosions

The energy of a Tsunami depends on its height (amplitude of the wave) and speed. So the total energy liberated will also depend on the number of waves that arrive at each location.

As a reference, under depths of four to five kilometers, the speed of the wave can go up to 600 kilometers per hour (372 miles per hour) and 800 kilometers per hour (500 miles per hour); with a crest separation larger than 100 kilometer. So, that the wave will keep (conserve) its energy until it breaks on the coast, it propagates very fast in the ocean and can, for example, cross the Pacific Ocean in less than one day.

When tsunamis approach the coast, they increase their height and are able to reach huge dimensions. As was seen in Chile, the damage can be observed in deaths; destruction of houses, buildings and ports; flooding of large amounts of land; and damage to transportation systems, energy networks, and water pipes. It's important to mention that objects dragged along by the tsunami may cause pollution and epidemic diseases in settlements.

There exists a major concentration of destructive tsunamis around the Pacific Basin. This is consequence of the successive earthquakes that are originated around what's known as the *fire belt*.

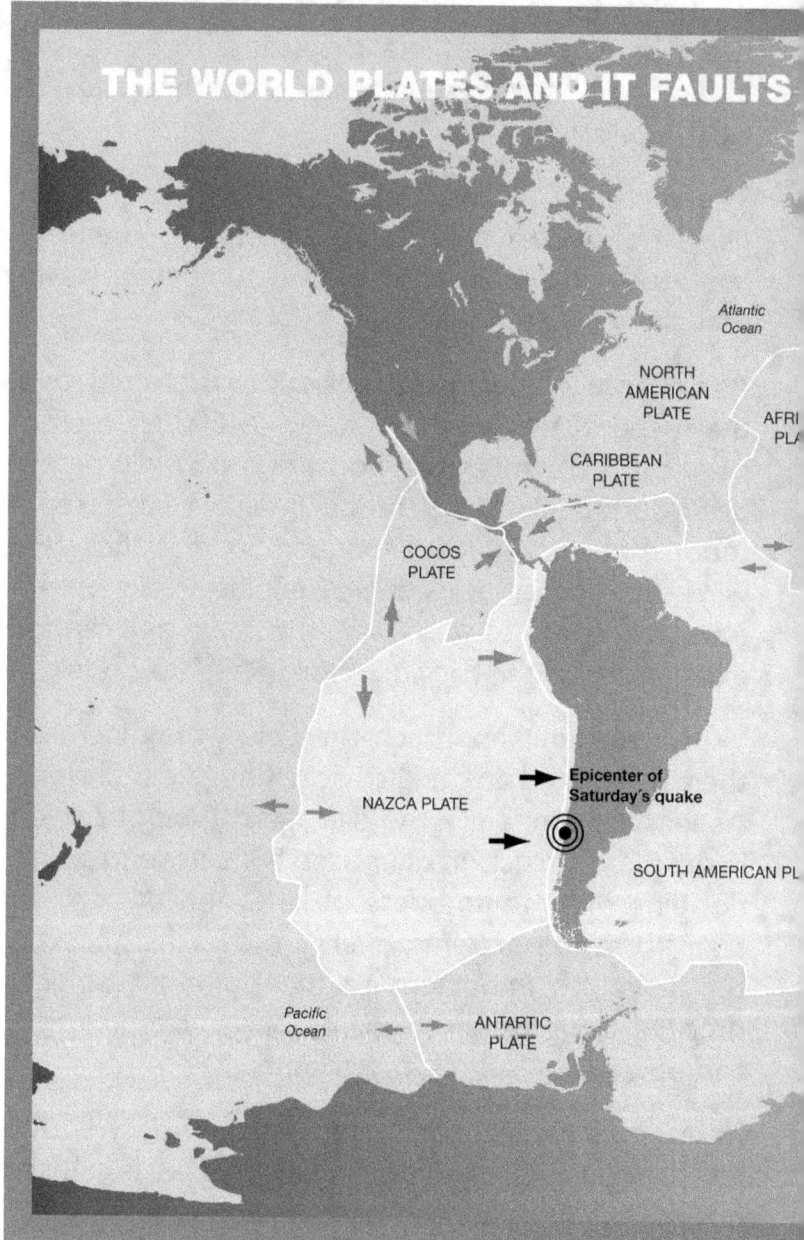

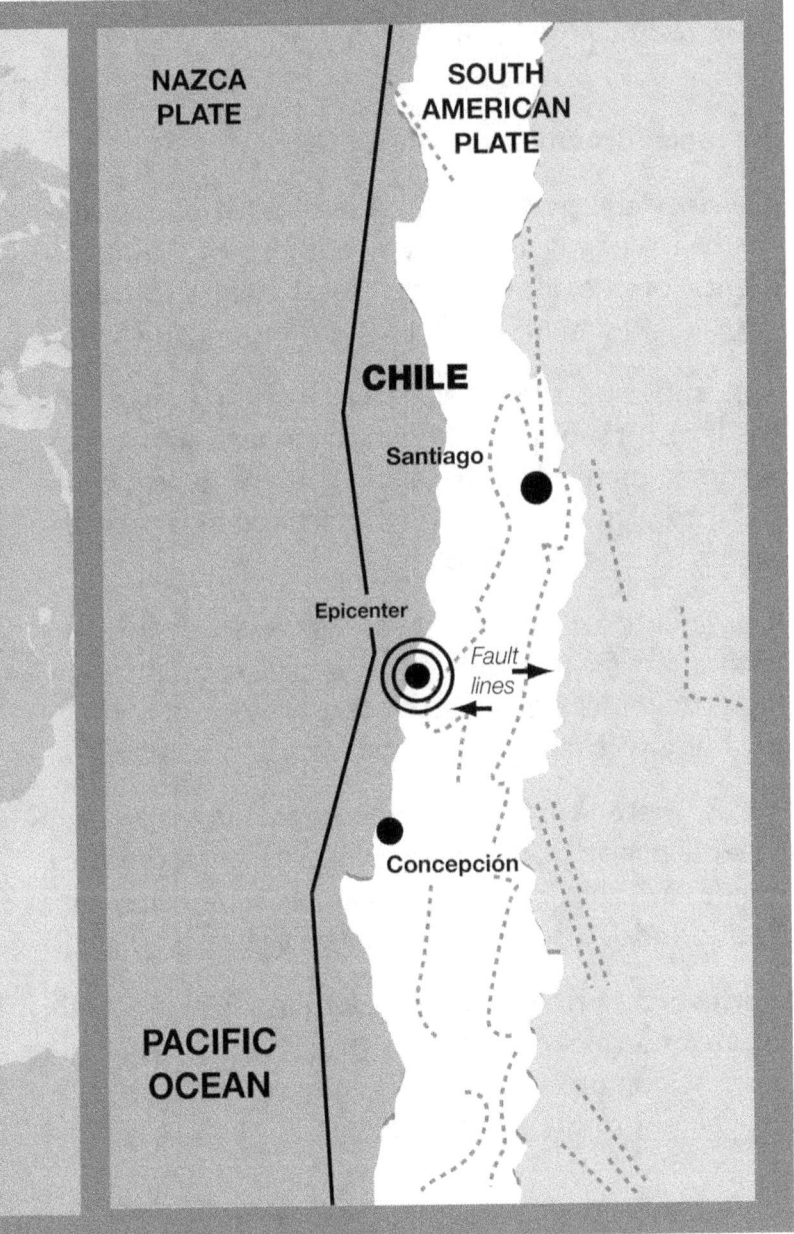

Since 1562, it is believed that the Chilean coast has been hit by more than thirty tsunamis.

Tectonic Movement

Thermal energy from the inner earth is not only the cause of volcanic activity but also of the movements of the lithospheric plates, which produce earthquakes and give life to great mountain chains.

Lithosphere is the most external and rigid layer of the earth. It's more rigid than the inner earth, which is more ductile and malleable due to its high temperature. This external shell is divided in large fragments called tectonic plates.

Continents are part of these plates, so they travel with them like a passenger. So we live over an ongoing puzzle formed of pieces that are not only moving but also changing in shape, size, and number.

Today, there exist seven main tectonic plates and several minor plates; the main plates are the Eurasian, African, South American, North American, Pacific, Indo Australian, and Antarctic plate.

Finally, the thermal energy located in the inner earth creates convection currents that move the materials that compose the earth's mantle and cause the movement of the tectonic plates.

Economic Shock in Simple Terms

Information elaborated on by the government indicates that overall damage (excluding human casualties) would reach the chilling amount of US$30,000 million. Considering that Chile's GDP is in the order of the US$160,000 million, the cost of this mega earthquake represents the 18.75 percent of the GDP.

Analyzing the damage by sector, it's possible to see that housing, and therefore the people, directly receive an important part of the shock. This is because the cost of reconstruction is has been estimated by the government in the following amounts classified by sectors:

Table 6: Economic Damages

Damages	
Sector	**Estimated Cost in U$ millions**
Housing	3,943
Education	3,015
Health	2,72
Energy	1,601
Infrastructure	1,725
Agriculture	601
Transport and Telecom	523
Municipalities	96
Industry and commerce	6,715
Reduced growth	7,606
Rubbles removal	1,117
Total	**29,662**

Source: Press information

Even though these huge numbers frighten, they don't reflect the urgency of the need; this region is in autumn, and the weather is growing rainy and cold day by day. Nor the urgency that people have for obtaining jobs and generating income: there are many settlements in which places of work don't exist anymore.

Think about those fishing boats that were left inland by the tsunami, or the many stores that were swept away by the sea.

As a consequence of this disaster, many companies have had to fire employees. About 6,000 people were fired in the first fourteen days after the earthquake, and by the end of March, 8,000 had lost their jobs. It's also important to consider that the workers whose jobs simply disappeared were fired invoking an article of the labor code that considers this situation a major force event; therefore there is no obligation of paying indemnifications.

Under some perspective damages are very high but under other, they are not so much considering that the seism affected badly several regions from V to VIII, including the metropolitan region, which is the most highly populated.

Table 7: Regional organization

Regional organization of Chile

	Region	Capital	Surface	Population
XV	Arica y Parinacota	Arica	16.873,3 km²	187.348
I	Tarapacá	Iquique	42.225,8 km²	300.301
II	Antofagasta	Antofagasta	126.049,1 km²	561.604
III	Atacama	Copiapó	75.176,2 km²	276.480
IV	Coquimbo	La Serena	40.579,9 km²	698.018
V	Valparaíso	Valparaíso	16.396,1 km²	1.720.588
RM	Metropolitana de Santiago	Santiago	15.403,2 km²	6.745.651
VI	Libertador General Bernardo O'Higgins	Rancagua	16.387,0 km²	866.249

VII	Maule	Talca	30.296,1 km²	991.542
VIII	Biobío	Concepción	37.068,7 km²	2.009.549
IX	La Araucanía	Temuco	31.842,3 km²	953.835
XIV	Los Ríos	Valdivia	18.429,5 km²	376.704
X	Los Lagos	Puerto Montt	48.583,6 km²	815.395
XI	Aisén de General Carlos Ibáñez del Campo	Coihaique	108.494,4 km²	102.632
XII	Magallanes y de la Antártica Chilena (I)	Punta Arenas	132.291,1 km²	157.574

Source: Wikipedia

Human Behavior in Simple Terms

The earthquake, has given us the opportunity to admire many good lessons of humanism, charity, generosity, and value.

After the earthquake, there were countless examples of people assuming roles of unknown heroes in front of others: seamen, policemen, soldiers, common citizens, medical service people, teachers, etc.

Among those, maybe rescue teams have been the most visible in the press. And they deserve special recognition because in their effort to find and rescue people from the rubble. They represent the heart of goodwill—people putting their lives at risk and working for several days, under very uncomfortable conditions, to help others.

There are so many cases of different outstanding people actions that an entire book could be written about them. However, I would note that they reflect important values or aspects of our society.

Real communication: After the earthquake people became more willing to talk to others. I'm not talking only about chat, twitter, or any indirect communication mechanism, even though they have been very useful after the emergency. I mean verbal communication with neighbors, colleagues, old friends, parents with who contact was lost or infrequent, and even strangers. I don't know the reason why, but it's clear to me that emergencies produce changes in the way human beings perceive their dependency on their communities and society. It seems like the lack of protection that we use to look for in our material goods makes us more human again, changing our willingness and even necessity to relate to others.

Solidarity: It has been very nice to see endless examples of solidarity among people. From people collecting clothes, food, and money, to others helping unknown citizens taking away rubbles. I'm not talking here about well-known international organizations like the Red Cross or local ones like the House of Christ. I'm talking about people who just decided to visit some of the most devastated areas, in some cases traveling hundreds of miles, just to contribute their effort, time, and willingness to help other unknown citizens.

Bravery: It is also encouraging to see how seamen got into their vessels, after securing their families, to try to put their vessels in a safe place. Just imagine what it would be like to head out to sea after an earthquake, when a tsunami is expected. In Talcahuano, brave seamen also took time to help other vessels set sail.

Of course, our society is not perfect, and so I would like to mention a case that surprised me. There was a woman at the hospital that had just given birth when the earthquake started. The obstetrician stopped the procedure while the nurses asked the mother to keep calm and hand the baby to the neonatologist. When the lights went off, the anesthesiologist fled. Once the earthquake stopped, the obstetrician finished the procedure, and fortunately, the baby was safe in the care of the neonatologist. But it result difficult to accept that anesthesiologist just abandoned her patient in such a difficult moment.

Generosity in Simple Terms

First Monday after the earthquake, several organizations started up initiatives to collect money. It's important to remember that after such an event, it takes some time to realize the dimension and scope of the tragedy. As the days went on, the Chilean people learned about the different realities that people were facing.

The result was good, and only one week after the earthquake, a huge twenty-four-hour "Teleton" was organized through which it was possible to collect:

a) US$ 42 million from people

b) US$ 49 million from companies

c) US$ 28 million from international governments and organizations

At this point it's very important to highlight that the international community helped in many different ways, including sending different supplies and also people. It was a *nice example of globalization.*

Edifications in Simple Terms

Chile has a strong construction code that was drafted to save lives. The idea of the code was to create structures that may get severely damaged but allow people to survive without being hurt.

Typical in Chile, and part of the code regulation, is the use of reinforced concrete structures because they are strong and flexible.

Of course, there was the building that fell down in Concepcion, killing several people inside. But even though it's terrible and is under investigation, the fact is that it's the *only* building built under the code force that collapsed.

This means that even though a large number of buildings in Concepción, Santiago, Viña del Mar, Talca or Rancagua were severely damaged, the number of people injured by these edifications is very low.

A fair question may be why there are buildings that have almost no structural damage while others are severely damaged? Well there are several experts studying these matters, and there may be a combination of different explanations, including the construction process, the soil mechanics, the possible existence of deep faults not detected by normal analysis, the building materials' quality, and the structural calculation.

There are some buildings that had problems related to the installation of pools or boilers on the roof of the building. In relation to the boilers, they suffered damage to their supports and fires. In both cases they were hard to stop because of the difficulty to reach them without elevators.

The earthquake raised the expectations of citizens about the damages suffered because of the flexibility of some structures. Until now it was common to have damage in nonstructural walls or dilatation joints. But after the seism, people not only want to live in a safe building or house, they also aspire to have minimal damage related to the movement, which is some-thing not addressed by construction code.

How Strong Was the Earthquake?

The Richter and Mercalli scales may be better for scientists and engineers. So I decided to include to lists to help "normal people" understand what happened, trying to compare this earthquakes with other ones that occurred in the past.

So, the following two lists show the largest earth-quakes ordered first by magnitude and then, the deadliest earthquakes ordered by fatalities. Unfor-tunately, the megaearthquake that gave rise to this book is classified in the seventh or fifth position, depending in how it's interpreted, in the list of larg-est earthquakes. But thanks to God, it doesn't ap-pear in the list of those causing the largest fatalities.

Table 8: Largest earthquakes by magnitude

Largest earthquakes by magnitude			
Pos.	Date	Location	Magnitude
1	May 22, 1960	Valdivia, Chile	9.5
2	December 26, 2004	Sumatra, Indonesia	9.3
3	March 27, 1964	Prince William Sound, Alaska, USA	9.2

4	November 4, 1952	Kamchatka, USSR	9.0
4	August 13, 1868	Arica, Chile	9.0
4	January 26, 1700	Cascadia subduction zone, Canada and USA	9.0
7	February 27, 2010	Maule, Chile	8.8
7	January 31, 1906	Ecuador–Colombia	8.8
7	November 25, 1833	Sumatra, Indonesia	8.8
10	February 4, 1965	Rat Islands, Alaska, USA	8.7
10	November 1, 1755	Lisbon, Kingdom of Portugal	8.7
10	July 8, 1730	Valparaiso, Chile	8.7
13	March 28, 2005	Sumatra, Indonesia	8.6
13	March 9, 1957	Andrean of Islands, Alaska, USA	8.6
13	August 15, 1950	Assam, India – Tibet, China	8.6
16	September 12, 2007	Sumatra, Indonesia	8.5
16	October 13, 1963	Kuril Islands	8.5
16	February 3, 1923	Kamchatka, USSR	8.5
16	November 11, 1922	Atacama Region, Chile	8.5
16	December 16, 1575	Valdivia, Kingdom of Chile	8.5

Source: Wikipedia

Table 9: Deadliest earthquakes on record

Rank	Name	Date	Location	Fatalities	Magnitude
		Deadliest earthquakes on record			
1	"Shaanxi"	01556-01-23 January 23, 1556	Shaanxi, China	830	8.0
2	"Tangshan"	01976-07-28 July 28, 1976	Tangshan, China	255	7.5
3	"Gansu"	01920-12-16 December 16, 1920	Ningxia–Gansu, China	234,117	7.8
4	"Haiti"	02010-01-12 January 12, 2010	Haiti	233	7.0
5	"Indian Ocean"	02004-12-26 December 26, 2004	Sumatra, Indonesia	230,21	9.3
6	"Aleppo"	01138-10-11 October 11, 1138	Aleppo, Syria	230	8.5
7	"Great Kant□"	01923-09-01 September 1, 1923	Kant□ region, Japan	142	7.9
8	"Ashgabat"	01948-10-06 October 6, 1948	Ashgabat, Turkmeni-stan	110	7.3

Source: Wikipedia

Other interesting information

Number of Earthquakes Worldwide for 2000-2009

The following chart was prepared by the USGD to show what I would call *the seismic signature* of the last decade. As it's possible to observe, there is shown a maximum casualties in 2004, terrifying result of the tsunami originated as a consequence of the earthquake of Indonesia on December 2004.

Table 10 Chart: world wide earthquake 2000-2009

		Number of Earthquakes		
Año	Estimated Deaths	Magnitud 6 to 6.9	Magnitud 7 to 7.9	Magnitud 8 to 8.9
2000	231	158	14	1
2001	21.357	126	15	1
2002	1.685	130	13	-
2003	33.819	140	14	1
2004	228.802	141	14	2
2005	82.364	140	10	1
2006	6.605	142	9	2
2007	712	178	14	4
2008	88.011	168	12	-
2009	1.787	142	16	1
Average	46.537	147	13	1

Source: USGS National Earthquake information Center

Personal Reflection

At this point, I would like to ask the readers to estimate the damage that a similar situation would have produced in your respective communities. Far from trying to exaggerate or create panic, an educated and prudent guess may be helpful in determining priorities for your survival and implementing a preparation plan.

I insist in the implied dilemma of these nature events, combined with the fact that they happened at intervals; it's easy to behave under extremes scenarios: either becoming paranoid thinking that we are close to a disaster or assuming that it'll never happen again.

CHAPTER 3

Lessons for the future

Main Thoughts

From what has been seen in this earthquake and the one that I experienced in 1985, damage may be quite different even in the same town. And this effect could be magnified in larger cities. Different reasons may be discovered for the large difference in damage: soil quality or the existence of a deep failure, construction quality, age of the neighborhood, height allowed in that part of the city, position relative to the coast or river in the case of tsunami, etc.

The good thing about this difference in damage is that if a person lives in a place that gets badly damaged, he or she may find help by moving to a parent's or friend's house within the same city. The bad thing about this is that the eventual lack of damage in one place does not imply that the rest of the city or properties behaved in the same way.

At least for the earthquake, not for tsunami, one of my conclusions is that managing an emergency becomes less complex and more effective, if people are prepared for emergencies and in this sense, keep some level of provisions ready to allow their families to survive for three to five days.

Another important impression that I have confirmed with this last earthquake is that a good way to anticipate complex situations is by creating scenarios, preparing courses of action, and then testing them in what I would call a recursive improving methodology.

The reality and our human nature have convinced me that it's hard to react correctly to unexpected situations. Especially if we don't have information about the event—something that takes time to obtain and process, and therefore is almost never available or complete in those difficult early moments of the emergency.

Under the risk of disapproval from who may believe that emergency reactions (in general) have been correctly addressed by governmental emergency organizations (maybe some politicians), I can only say that evidence of consequences of large disasters makes me disagree. Take for example situations like the tsunami in Indonesia in December 2004, with more than 200,000 casualties or the megadisaster caused by the Hurricane Katrina, in New Orleans in 2005, taking more than 1,800 human lives and costing US$75,000 million. In each case, clear information about the situation was obtained after a significant delay and therefore resources assigned with the limitations and restrictions proper of that reality.

In the case of the megaearthquake in Chile, it is clear that all the authorities, politicians, militaries, institutions, and people involved in managing the emergency did their best including mistakes, which are also part of our human nature and lack of information. But even though their good intentions helped a lot, they weren't enough and several people died. In relation to this last, my personal opinion is that legalizing the problems, by charging people with heated demands, represents a non-conducive and non-constructive behavior.

So once again, after emphasizing our human limitations to manage large emergencies, assuming a more humble attitude and therefore not blaming people or institutions for the casualties of this event, I'll try to create a framework of analysis, with the idea that could help people who live in areas with earthquake or tsunamis potential.

Trying to simplify something very complex, I see that one analysis could be run grouping the preparation level under three categories:

a) Matters in which the home work was done and so, the results may not be perfect but were acceptable considering such a strong strike from nature.

b) Matters considered by some people but that weren't correctly coordinated, trained, or tested and

c) Matters in which there was no preparation at all.

a) First category: home work done

Construction code: As was explained earlier in this book, it prevented buildings, with the exception of one, from collapsing, which saved thousands of lives. This is a major achievement when you consider that Chile is not a developed nation and still has a high level of poverty.

Military and naval forces: it's amazing to realize, the outstanding preparation, capability, and flexibility shown by the Chilean military forces. To show this, let me give just some brief examples. The navy sent a frigate to Juan Fernandez Island with provisions and medical personnel only a few hours after the tsunami damage was realized. Hundreds of homeless people from Talcahuano were housed at the marine headquarters, located in the same devastated city. The air force had airplanes ready only two hours after the earthquake and only a few hours later deployed a large operation with helicopters giving support to the more damaged areas. The army was ready not only to keep the order and maintain the safety of the civilians, but also to deploy emergency medical services, again only hours after the occurrence of the disaster. Unfortunately, the emergency rolls of the armed forces that were in force since 1977, was changed and reduced in favor of the ONEMI (National emergency office), on March 12, 2002.

b) Second category: incomplete preparation

There existed a National Emergency Office (ONE-MI), from which the central government were supposed to coordinate the resources that were sent to different areas. Even though this office is under evaluation and has a lot of people questioning the effectiveness of its actions, the fact is that it worked as a command point for the president of the country and the ministers, also facilitating some level of coordination with resources from the army, navy, air force, police and other institutions.

In relation to this last, I'm convinced that there existed two main problems: a) there existed a structural design one, in the way the ONEMI should direct resources that they didn't fully understand (like military capabilities, for example) and based on information that was difficult (or impossible) to obtain b) Political authorities confused their hierarchy with an operative authority, as they began trying to give direct orders, reducing the potential of the ONEMI (national emergency office) as well as the capability of processing data in order to obtain valuable information.

Tsunami: The SHOA (hydrographic and oceanographic service of the Chilean navy) prepared for years with flooding scenarios and implemented a system oriented to detect and react in these events. After the experience in Indonesia in December

2004, signs were placed in coastal cities to indicate an evacuation route and showing where to go in case of a tsunami alert. Although it's currently under investigation, available information indicates that the navy communicated the alert by radio to ONEMI, activating the tsunami alert only seventeen minutes after the earthquake and by fax, thirty-two minutes after the event. For reasons still not completely understood and may be related to the damage that the earthquake caused to communications systems, the alert didn't reach all the coastal residents; many people had no warning. But the problem didn't stop there; it seems that SHOA received incorrect data from buoys indicating that there was no variation in the sea levels and so canceled the tsunami alert by 4:56 A.M., before the arrival of other waves.

The reality indicates that the communications were an issue not yet resolved and that the speed at which the tsunami advanced made it difficult to protect people living within what is called by some expert as the "sacrifice area". Therefore, people should be trained to react without waiting for communication from any emergency office at all.

Implementation of the construction code: The supervision of the implementation of the construction code could be in this category because there exist no formal supervision of the construction companies. I'm not the person to suggest what the best solution is, but it may be related to a construction quality certification.

Insurances: When people bought using credit, it was mandatory to obtain fire insurance that in some cases may have included additional earthquake coverage. My opinion is that insurance should be mandatory for every property owner and should cover the total amount of the property value and may have different options with regard to deductibles. It would be also important, from a customer's point of view, to have similar or equalized offers in order to let common people compare what may become sophisticated (noncomparable) instruments.

Emergency rescue teams: When realizing that there were people trapped under rubble I can only think that time means life. Emergency teams must arrive as soon as possible. Also, they need to know which areas are the most damaged, where there are more likely to be survivors, and ideally, have the blueprints of the structures in which they are going to work. These teams are composed of people who also have to be supported in terms of food, water, and housing.

Security for damaged communities: There was no special security established after the tragedy, and some people reacted in a violent way stealing all that they could. Because of this, more than thirty hours after the disaster, the government sent the military to take control of the area.

Communication networks: Even though Chile has an extensive mobile- and fixed-phone network that works pretty well under normal conditions, it was almost impossible to establish phone communications for up to several hours after the earthquake. The lack of communication was a bad thing. In prevented announcing emergencies, generated uncertainty among citizens, and also increased the number of people traveling to ascertain the state of their families. This doesn't mean that the system failed; it's just related to the capacity to manage an overwhelming demand for phone communications required during an emergency and to do it without an emergency source of energy. After this experience, it would be wise to explore ways to ensure mobile networks can accommodate heavier demand during emergencies. Also, priority should be given to emergency services and to training users to prefer messaging services instead of voice communications.

Water: Several water ducts were damaged in many places. The more affected areas received water only by trucks because the damage made impossible to reestablish the service for three weeks after the earthquake.

Electricity: Chile has two main (not redundant) electric transmission systems—the SIC and the SING. The latter is located in the north of the country. SIC supplies consumers from Taltal down to Chiloe, which may mean nothing to people who don't live in Chile, but the following data may make it more meaningful: SIC serves 93 percent of the population of the country: approximately twelve million people. The earthquake affected projects under construction to deliver 950 mega watts, which corresponds to the 57 percent of all the plants under development.

Roads: Several roads, bridges, and highways collapsed or suffered heavy damage. There are some cases in which vehicles that were traveling over these roads suffered accidents in which their occupants were injured.

c) Third Category: No Preparation

We were not rigorous enough to prevent construction of cities and settlements in places vulnerable to tsunamis. And I say *rigorous* instead of *visionary* because tsunamis are natural events that Chilean coast suffered in the past.

Food for victims: The information known up to this moment indicates that some areas hit by the earthquake and more specifically by the tsunami, were without food and medicines for a few days. The roads were damaged; the stores, supermarkets and the employees were affected also. So some people -especially parents living in the sacrifice area- went through uncertain days trying to obtain food and water.

Hospitals from Talca to the southern Concepción suffered great damage. Even though the idea is that this kind of (public) edifications should be able to get through a 9.0 earthquake, the reality showed that for different reasons, including the building's age, many of them become inoperative with the seism. Field hospitals partially compensated for this situation.

Elevators: There is no practice or code enforcement for equipping elevators with seismic sensors and emergency stop mechanisms. In a country were intense earthquakes are common, my belief is that this would help people by eliminating the traumatic experience of being trapped inside an elevator.

Old structures revision: Maybe because of the cost, but we had no effective plan for checking old structures.

Preset tsunami escape areas: Some cities prepared signaling, but it lacked accurate indications of places to go and depended on the relative position within the city. Also, people tended to move by car, causing congestion and with that, making evacuation unsafe.

Logistic and support for rescue teams: Volunteers and rescue teams need to be supported to make them more efficient. Basic elements like water, food, lodging, and bathrooms should be considered in any good emergency plan.

Food for affected people: When the damage caused by natural disasters is too intense, there may be people that lose all their provisions and their ability to obtain food. It's important to coordinate mechanisms to distribute food; this could be done either with the cooperation of the army, the coordination the supermarket chains, etc.

Posts in the cities: I have never liked the electric and communication cables that run a few meters over the streets of Chile, hanging from posts all around each and every city. I consider them dangerous, ugly, and contrary to good quality of life. As it can be imagined, they are not a safe method of conducting cables, even though they are cheaper than using underground ducts because they may get broken by the earthquake or damaged by a tree, a car, or other element that may hit them.

Government reaction

Only twelve days after the megaearthquake, Chile had a change of government and also of ruling coalition. Under no political influence, I would say that the new government reacted quickly doing a fast diagnosis and then creating a three-phase plan:

a) Immediate Emergency: Oriented to assure the provision of food, water, and shelter; re-establishment of basic services; and public order. It was created by an emergency committee that works with military forces (navy, army, and air force) in conjunction with different government authorities.

b) Winter emergency: Directed by the Planning Ministry and, under close coordination with municipalities, is oriented to supervise the delivery of 40,000 emergency houses, as well as 60,000 hiring incentives, and other legal and administrative initiatives.

c) Reconstruction: Coordinated by an inter-ministry committee, it has the mission to create a program consistent with the vision of a better country.

Prepare yourself and family—do your homework

You will never be ready for a megaearthquake, it will always be inopportune. So based on my personal experience with two strong earthquakes, I would suggest going through the following preventive actions:

- Insurance: Contract insurance to cover your properties against probable events. Think about fires, earthquakes, tsunamis, volcanic eruptions, tornados, hurricanes, floods, storms, etc. Obtain advice from an insurance expert, but do it now. Take into consideration that if there are events with low probability of occurrence like tsunamis for example, it doesn't matter; the prime should have a low cost.

- Depending on where you live and where you typically go, prepare a brief plan of alternative places to which you may evacuate. If you live near the coast, verify the nearest point where you would be safe in a tsunami event. Verify that these places and the route to reach them are safe. This will allow you to relax because in the event of an emergency, you will remember what to do.

- If you have to leave your home after an earthquake, be extremely careful on the road:

  o After these events, the roads and also bridges may be damaged. Also it's well known that aftershocks follow earthquakes, and while they typically have less intensity, they can also be very strong.

  o Select a road that is not at risk of being hit by a tsunami or avalanche.

- At home:

  o Verify that your house was built according to anti-seismic specifications.

  o Check that it's not built over an area that could get hit by a tsunami.

  o Equip your home with emergency lights and fire-control elements.

  o Have, candles, matches, and a good lantern with spare batteries.

  o Consider always having extra bottles of water and use a *fi-fo* (first in-first out) consuming system.

  o Buy a small battery radio and keep a new battery pack.

  o Keep extra nonperishable food in your pantry and consume them under a fi-fo mode; you should always have food for two to three days.

  o If you use medicine, consider having a two-day reserve.

  o If you have a car, keep it operative and with enough gas to travel to a safe place.

  o Keep always some cash on hand.

o Leave corridors free of obstacles at home, especially at night.

o If you live in an earthquake area, take care of securing large furniture so they can't hit a person.

o Have beds clear of anything heavy that may hit someone.

o It would be ideal to have camping equipment correctly stored and ready to go.

o If you have a pool, consider having a system able to feed water for shower and water closet.

o Have fire extinguishers ready to use in case of fire and train your family in using them.

- If you live in a building, consider checking if there are emergency lighting in stairways and corridors, and fire control systems. Exits should be properly operative and marked.

- Evaluate curses of actions in case members of the family are out of home: working, at school, in a party, etc.

- Train your family and friends in these simple but useful practices.

Final words

The mega earthquake that hit Chile was a very damaging event. Fortunately the number of casualties is estimated at less than 500, which is still too high; but it's also a low number if it's compared to other similar events that occurred in other countries.

Reasons for these "good" results may be found in the experience learned from the past (for example the largest earthquake ever measured: 9.5 in 1960), which motivated the actual construction code and also transmitted the custom for coastal residents of moving to a hill after an intense earthquake, etc.

But one of the problems that I see is that some people don't understand or accept the implications of such an intense earthquake. It seems like the expectation was to have no damage at all. There have been a large number of lawsuits initiated, seeking those responsible for the damage or even for the death of the victims.

My feeling is that these suits show, at least, an incomplete preparation and more than that: the role of justice has been misunderstood. I'm not sure if the media influenced this litigious behavior by showing dramatic images of the earthquake, suggesting bad construction practices or insinuating a confuse behavior of authorities, etc. But it's my belief that it's important to realize that a megaearthquake like this, is a catastrophic event that creates situations that exceed human capabilities and produce unwanted results.

I really wish that there had been no casualties or economic losses, but I don't agree with demanding authorities, construction companies, navy, emergency office, etc. take legal responsibility. In contrast, it's clear to me that people did their best and if it didn't work, it was due to structural failures of the ONEMI organization, human limitations and some wrong short term political incentives, rather than negligence as has been suggested by some.

With the experience of this event, if I had the chance to step back to January 2010, I would have improved the emergency office capabilities for monitoring tsunamis; reminded people that tsunamis may be composed of several waves and that the danger may exist during several hours; include property insurance in the payment of property taxes, and increase the emergency capabilities of mobile phone companies.

Now, if I could step back a bit further, I would have also like to perform the following actions: re-structure an emergency plan, defining clear rolls for each and every institution and also a permanent training program; improve the construction code in order to make buildings even more safe; make earthquake insurance obligatory for all properties, again maybe by including in the property taxes; create a tax incentive program for replacing old structures (especially those of adobe); improve communications systems; implement a ready reserve of emergency materials such as houses, hospitals, mechanical bridges; create a VHF and HF communication network; prearrange an emergency food delivery agreement with supermarkets; improve international cooperation and learning.

In relation to the new emergency organization, I would encourage the difference between the high level authorities' roll, as responsible for taking care of the correct execution of the plan and the people responsible for commanding the resources: the emergency office. The first need to receive already processed information (especially president and ministers), etc. but is the emergency office staff, the one that have to coordinate the resources. Not understanding this would be like having the US President flying (guiding) the Air-force one, just because he is the highest ranked on board.

Looking forward in Chile, we should not only take care of the reconstruction, but also try to reduce potential damage in what could be our next bad scenario: an earthquake and tsunami in northern Chile. To that end, I would like to ask experts around the world for mechanisms that may reduce the intensity of a tsunami and therefore, improve the protection of people and edifications built along the coast—at least, for the largest cities located in northern Chile: Arica, Iquique, and Antofagasta.

Finally, I really hope that this book may help some people living along the Pacific RIM, as a reference in order to review their local situation and improve the preparation for eventual future emergencies.

COMPLEMENTARY INFORMATION

If you are interested in learning more about these devastating events, there are organizations that offer interesting information through the World Wide Web.

1) The NOAA has valuable information. Some very interesting links are the following:

   - http://nctr.pmel.noaa.gov/ chile20100227/20100227Chile.mov

   - http://nctr.pmel.noaa.gov/chile20100227/

2) The International Tsunami information Center has valuable information and makes it available in several languages. http://ioc3.unesco.org/itic/ contents.php?id=216

3) Complementary information for earthquake and potential tsunami area residents

   The U.S. Geological Service has created an excellent, easy-to-use Web site where information related to different hazards, including the earthquakes, is available within only a few minutes of the event. http://www.usgs.gov/

4) International oceanographic commission http://ioc-unesco.org/

The report "Five Years After: The Tsunami in the Indian Ocean, From strategy to implementation" is available at http://ioc-unesco.org/index.php?option=com_oe&task=viewDocumentRecord&docID=4649

5) Complementary information for California residents

This is an interesting Web site that includes Estimation of Future Earthquake Losses in California, prepared by B. Rowshandel, M. Reichle, C. Wills, T. Cao, M. Petersen(*), D. Branum, and J. Davis (California Geological Survey) http://www.consrv.ca.gov/cgs/rghm/loss/Pages/index.aspx

Quoting the first part of the executive summary, "Using the latest information on earthquake hazard in California and the publicly available demographic data, we have made estimations of expected future earthquake economic losses in the State. The estimates presented in this paper are for two categories: scenario earthquake loss, and annualized earthquake loss."

ACKNOWLEDGMENTS

This book was possible because of the help of many people, so even at risk of forgetting someone, I would like to thank the following people, in no particular or merit order: my wife, who always kept reminding me that humans need to sleep and who looked for interesting press articles.

My children, Jose Tomas and Maria Trinidad, for helping me to find good, easy to understand geological information in their science textbooks.

Professor and structural engineer Patricio Bonelli, who organized a group of experts with whom I had the opportunity to visit damaged buildings to see how structures behaved during an earthquake and also to seek reparation alternatives that could save buildings from demolition.

The NOAA / PMEL / Center for Tsunami Research

My uncles Angelica, Pablo, and Jorge Pinochet, who despite being in the United States, Mexico, and Spain, respectively, helped with corrections and ideas.

To Robert Ellias, who from the USFCA helped me find editors and publishers.

To Sally and Bayne, who helped with corrections and suggestions.

Ocean engineer and tsunami expert Ms. C. Patricio Winckler, whose research helped me better understand the way the sea behaves.

My cousin Felipe Gazitua, who helped in every way I requested.

To Mr. Luis Saldarriaga, who can be reached at http://www.exobiologia.8m.com/main.html for his help with the images and maps.